THE TORTOISE FARMER'S HANDBOOK

From Hatchlings To Healthy Adults: Expert
Insights On Breeding, Rearing, And Ensuring
The Well-Being Of Your Pet Companion

PRESTON DAMIEN

Contents

DISCLAIMER

The information provided in this book is intended for general informational and educational purposes only. The author of this book is not engaged in rendering professional or veterinary advice. The content in this book is based on the author's personal experiences, research, and knowledge, and it is not a substitute for professional advice.

The practices and techniques described in this book are meant to provide a foundation for understanding, but specific circumstances may require individualized approaches. It is recommended that readers consult with experienced veterinarians for guidance.

The author and publisher do not guarantee the accuracy, completeness, or timeliness of the information presented in this book. The reader is responsible for ensuring that the practices and recommendations align with current laws, regulations, and safety standards in their respective locations.

By reading this book, you agree that the author will not be held responsible for any actions you take based on the information presented in this book. You are solely responsible for any consequences or outcomes resulting from your farming endeavors. Always exercise caution and seek professional advice when necessary to ensure the well-being of your farm animals and to comply with all relevant laws and regulations.

CHAPTER ONE

Tortoise Farming: An Overview

Tortoise farming is the activity of breeding and maintaining tortoises for a variety of objectives such as conservation, pet commerce, research, and even culinary use. Tortoises are slow-growing reptiles noted for their longevity, making them a fascinating and one-of-a-kind farming alternative. The growing demand for tortoises, both as pets and for traditional medicine, has drawn attention to this business.

Why Do People Raise Tortoises?

Tortoise farming has numerous functions. For starters, it can help relieve pressure on wild tortoise populations by providing a sustainable source of these animals for the pet sector and other businesses. Furthermore, it enables regulated breeding operations, which contribute to the conservation of endangered tortoise species. Tortoises are also treasured for their flesh, which is considered a delicacy in some cultures. Their cultivation for culinary reasons has the potential to minimize the illicit traffic of wild-caught tortoises while also encouraging ethical and environmental techniques.

Tortoise Breeds For Farming

Several tortoise species, including the popular African spurred tortoise, Russian tortoise, and Aldabra giant tortoise, are ideal for farming. Climate, area, and market demand all influence species selection. Because different species have different care needs and growth rates, prospective tortoise farmers must carefully evaluate these issues.

Considerations For Legal And Ethical Compliance

Tortoise farming is governed by laws and ethical issues. Many nations have rules in place that restrict the breeding, sale, and ownership of tortoises, to protect both the creatures and the environment. Ensuring adequate care, habitat conditions, and humane treatment of farmed tortoises are all ethical problems. Conservation efforts should focus on preserving genetic variety and minimizing disease introduction. To preserve the long-term well-being of these interesting reptiles, farming should always be done with sustainability and animal welfare in mind.

Selecting The Best Tortoise Species

When starting a turtle farm, it is critical to choose the right tortoise species. The selection should be in line with your objectives, resources, and local climate. Each species has distinct care needs and development trends. The Russian tortoise, Greek tortoise, and sulcata tortoise are also popular alternatives.

The Russian tortoise, for example, is a smaller variety that is ideal for individuals who have little room. Its small stature implies cheaper feed and enclosure expenses. The sulcata tortoise, on the other hand, develops significantly larger and requires a lot of room and resources.

Consider your ability to provide long-term care, since tortoises may live for decades.

Farming Tortoise Popular Species

Tortoise farmers frequently choose species that adapt well to captivity and have a commercial demand. Russian and Greek tortoises are popular owing to their small size and kind temperament. They flourish in captivity, making them an excellent choice for novices. Furthermore, their reduced stature necessitates less room and food.

Although bigger and more difficult to care for, the sulcata tortoise is well-known for its toughness. Because they are frequently sold as pets or for breeding, their increased size equates to better profit margins.

Species Selection Considerations

Consider your climate, available area, and long-term commitment before selecting a tortoise species for farming. Investigate their nutritional requirements, habitat preferences, and growth rates. Consider local laws governing exotic pet ownership and trade.

Consider your target market as well. Some tortoise species are more popular as pets, while others are prized for their flesh or as breeding stock. Choose a species that fits your objectives and resources.

Accessibility And Sourcing

Make sure you have a dependable supplier for tortoises. Investigate local breeders, reliable internet sellers, and conservation groups. Buying tortoises in the wild can contribute to habitat damage and population decrease.

Before purchasing tortoises, check their legal status in your region and receive the required licenses. Veterinarians with tortoise knowledge might help locate healthy individuals. Remember that the availability of various species varies by location, so do your homework before making your choice.

Creating An Appropriate Environment For Tortoise Farming

When it comes to turtle farming, creating an appropriate ecosystem is critical. Tortoises must survive in an environment that closely resembles their native habitat. This includes providing a large, safe habitat with enough sunlight, good temperature regulation, and a diverse feed. To recreate their original topography, the habitat should have a mix of sandy, grassy, and rocky regions.

Environmental Concerns In Tortoise Farming

Environmental concerns are critical in turtle farming. Tortoises are highly sensitive to environmental changes, and pollutants, insecticides, and invasive species can all have a negative influence. Farmers must keep their tortoises in a clean, chemical-free environment to guarantee their health. Furthermore, protecting the natural habitat and reducing human disturbance in the region is critical for tortoise farming's long-term viability.

Tortoise Farming Temperature And Lighting

In tortoise farming, maintaining the proper temperature and lighting conditions is critical. Tortoises are ectothermic, which means they rely on outside heat to control their body temperature. Farmers must offer access to basking spots where tortoises may soak up the warmth of the sun. Proper illumination is also necessary for calcium metabolism and shell health, including UVB exposure. These elements must be monitored and controlled to prevent health problems in confined tortoises.

Tortoise Shelter And Hideouts In Farming

Tortoises require shelter and hiding places to defend themselves from adverse weather and predators. It is critical to provide natural or manmade shelters such as burrows, logs, or hiding boxes. These safe zones provide tortoises with a sense of security and aid in the regulation of their body temperature. In addition to reducing stress and promoting healthy behavior in confined tortoises, adequate housing ensures their general well-being on the farm.

CHAPTER TWO

Tortoise Farming Feeding And Nutrition

Tortoise farming necessitates close attention to feeding and nutrition to preserve the health and lifespan of these reptiles. A healthy diet that includes a range of leafy greens, vegetables, and fruits is essential. Increasing calcium consumption is critical for shell health, which is frequently accomplished through supplements or the availability of calcium-rich foods. Since tortoises are herbivores, high-protein and high-fat foods should be avoided in their diet. It is important to do a study and adjust their food to their species, taking into account parameters like as age and size. Regularly monitoring their weight and health indicators allows them to change their diet as needed.

Adding To Their Diet:

It is critical to supplement a tortoise's food to suit specialized nutritional demands.

Calcium supplements are widely utilized, particularly in species that require greater amounts for shell growth. Vitamin supplements, such as vitamin D3, may also be required, especially for tortoises kept inside with little natural sunshine exposure. However, striking a balance is critical, since over-supplementation can lead to health problems. Always visit a veterinarian or reptile nutritionist to identify the best nutrients for your tortoise's species and unique needs.

Common Tortoise Farming Mistakes To Avoid:

Tortoise farming necessitates careful planning to prevent typical blunders that might jeopardize the health of these interesting creatures. One common blunder is providing an unbalanced or improper diet.

High-protein or low-fiber diets, as well as an abundance of fruits, can contribute to digestive issues and obesity.

Common problems include poor living circumstances, such as limited room or incorrect temperature and humidity levels. Neglecting veterinary treatment, especially periodic check-ups, is another error that can lead to undiagnosed health problems.

Finally, misinterpreting a tortoise's behavior or wants might result in stress and disease; it's critical to learn and understand the tortoise species in question. Education and attentiveness are essential for effective tortoise farming.

Tortoise Farming Health And Veterinary Care

Tortoise farming necessitates meticulous health and veterinary care to maintain the well-being of these amazing reptiles.

Regular examinations and attention to their requirements are critical to their lifespan and quality of life. Here are some important features of tortoise health and veterinarian care:

Tortoise Health Indicators:

1. Tortoises in good health are generally active and alert. They should wander around their enclosure, investigate, and show an interest in what they see.

2. A healthy tortoise will have a steady and robust appetite. They should avidly consume a range of leafy greens, vegetables, and fruits regularly.

3. Tortoises should keep a constant weight within the standards of their species. Weight loss or increase might be an indication of a health problem.

4. Clear Eyes and Nose: They should have clear eyes and nasal passageways with no discharge or edema.

5. Smooth Shell: The shell should be smooth and free of fractures and irregularities.

6. Regular Bowel Movements: Tortoises in good health have regular bowel movements.

7. Breathing Ease: They should be able to breathe normally without any wheezing or strained breathing.

Common Health Concerns:

1. Respiratory diseases: Tortoises are susceptible to respiratory diseases, which are frequently caused by insufficient temperature or humidity conditions.

2. Metabolic Bone Disease: A poor diet or inadequate illumination can result in weaker bones and abnormalities.

3. Internal and external parasites can harm tortoises if they are not treated immediately.

4. Shell Rot: Shell infections can be caused by poor hygiene and moist circumstances.

5. Infections of the Eyes and Mouth: Infections of the eyes or mouth can arise, causing difficulties eating and visual impairments.

Routine Health Checks: Inspect your tortoises regularly for any symptoms of disease or anomalies. Keep an eye on their weight, behavior, and hunger. Keep enclosures clean and at the proper temperature and humidity conditions. To avoid health problems, provide a balanced diet and UVB lighting.

Locating a Reptile Veterinarian: Look for a veterinarian who has expertise in treating reptiles, particularly tortoises. They should be aware of their specific requirements and have access to diagnostic instruments and treatments. For advice on trustworthy reptile veterinarians in your region, contact local herpetological groups or internet forums. Regular veterinary visits are essential in tortoise farming for proactive health care.

Tortoises For Breeding

Breeding tortoises is a time-consuming procedure that involves thorough preparation and knowledge of these reptiles' particular biology. Tortoises achieve sexual maturity at different ages depending on the species, with some taking several years to a few decades.

A regulated setting with adequate temperature, lighting, and habitat conditions is generally required for successful breeding.

Reproductive Habits:

Tortoises have diverse reproductive practices that differ between species. Male tortoises frequently perform courting behaviors such as head bobbing, biting, and pursuing the female. Females often deposit eggs within a few weeks of mating. Understanding these actions is critical for effective reproduction and monitoring.

Breeding Pair Selection:

It is critical to select the best breeding pairings to produce healthy progeny. Genetic variety, compatibility, and health are all factors to consider.

To minimize genetic problems in the progeny, inbreeding should be avoided. Breeding programs frequently promote genetic variety to protect the long-term health of captive populations.

Incubation And Nesting:

Tortoise eggs are usually hidden in nests, and incubation times vary according to species and environmental circumstances.

It is critical to provide a good nesting place with sufficient substrate and temperature control. The baby tortoises that emerge from the eggs are delicate and require careful care to ensure their survival.

Taking Care Of Hatchlings:

Hatchling tortoises are delicate and must be kept in a regulated setting to keep predators and environmental risks at bay. During this important stage of growth, proper nutrition, temperature, and humidity are critical.

As hatchlings mature, they are progressively introduced to their native environment, with careful monitoring to guarantee their well-being. Successful tortoise farming includes not only breeding but also the care and protection of these amazing reptiles.

CHAPTER THREE

Legal And Ethical Issues In Tortoise Farming

Tortoise farming, like any other kind of animal husbandry, imposes substantial legal and ethical obligations. Farmers are required by law to follow local, state, and national restrictions controlling the breeding, selling, and care of tortoises. Obtaining permissions and licenses, providing suitable cages and veterinary care, and adhering to species-specific rules are all part of this.

In tortoise farming, ethical issues go beyond basic legality. Farmers must emphasize animal welfare by providing proper shelter, feeding, and medical treatment. Breeding tortoises should also adhere to ethical standards, preventing inbreeding and preserving genetic variety within captive populations.

Furthermore, the ethical aspects of turtle farming have an impact on conservation efforts. Because many tortoise species are endangered or vulnerable, good breeding methods are critical to their survival. Farms should work with conservation groups, exchange expertise, and contribute to research to help conserve and recover these species.

Tortoise Farming Regulations And Permits

Tortoise farming is governed by a tangle of rules and regulations. These restrictions differ depending on the region and the species being farmed. Tortoise farmers often require licenses to purchase, breed, and sell tortoises, particularly if the species involved is protected under wildlife protection regulations.

Minimum specifications for tortoise enclosures are frequently specified by regulations, which must offer enough space, temperature, and predator protection.

Veterinary care and health inspections may also be required to ensure the animals' well-being.

International legislation, such as the Convention on International Trading in Endangered Species of Wild Fauna and Flora (CITES), may limit the cross-border trading of certain tortoise species in particular situations. These restrictions must be followed to prevent illicit trafficking and safeguard endangered animals.

Tortoise farmers must be educated about changing legislation, keep proper records, and collaborate closely with appropriate authorities to ensure they meet all legal criteria.

Tortoise Farming Conservation And Ethical Practices

Conservation is critical in tortoise farming. Many tortoise species are vulnerable or endangered as a result of habitat degradation and illegal wildlife trading. Tortoise farming ethics must prioritize the protection of these animals.

Farmers should prioritize breeding programs that enhance genetic variety and limit the risk of inbreeding, and they should strive to release captive-bred tortoises into their natural habitats wherever feasible. Collaboration with conservation groups is critical for knowledge sharing and contributing to larger conservation initiatives.

Furthermore, ethical tortoise farming procedures emphasize education and awareness. Farmers should educate their consumers about proper pet ownership and tortoise needs. It is also critical for conservation to discourage the illegal trade in wild-caught tortoises.

In tortoise farming, conservation, and ethics work hand in hand, with appropriate breeding and care techniques acting as a critical pillar in the preservation of these magnificent reptiles.

The Morality Of Selling Tortoises

The ethical implications of selling tortoises lie in responsible ownership, transparency, and animal care. Sellers must ensure that prospective purchasers understand the long-term commitment and particular care requirements of owning a tortoise as a pet.

It is immoral to sell tortoises to anyone who does not have the expertise or means to care for them properly. Sellers must do full background checks on customers, inquire about their experience with reptiles, and provide advice on correct husbandry procedures.

Transparency is essential in ethical tortoise sales. Sellers must provide correct information about the species, age, and health of the tortoises they are selling. To guarantee that the new owner can satisfy these criteria, they must supply detailed information on the tortoise's history, food, and housing needs.

Ethical retailers also avoid participating in the illicit wildlife trade by ensuring that the tortoises they offer are lawful. They should have the necessary permissions and papers to prove the tortoises' legal status.

Finally, the ethics of selling tortoises lay in improving animal welfare and encouraging appropriate ownership habits.

Tortoises For Sale And Marketing

Tortoise farming entails more than simply breeding and nurturing these one-of-a-kind reptiles; it also needs strong marketing and sales methods. Tortoise marketing is a specialized undertaking that caters to reptile aficionados, pet owners, and environmentalists. Here are some important factors to consider while marketing and selling tortoises:

1. Target Audience: Determine who your target audience is. Are you selling them as pets, for conservation, or both? Understanding your target market allows you to adjust your marketing strategy.

2. Create a professional website or use internet markets to display your tortoises.

To attract potential customers, high-quality images and thorough descriptions are required.

3. Connect with reptile forums, social media groups, and local reptile clubs to network. Engage the community to establish trust and confidence.

4. Permits and requirements: Ensure that all local and international requirements for selling tortoises are followed. Obtain all required permissions and licenses.

5. Educational Materials: Distribute educational materials regarding tortoises, their care, and conservation initiatives. This may entice potential consumers who are enthusiastic about these reptiles.

6. Customer Service: Provide great customer service to handle inquiries and complaints as soon as possible. A positive reputation is essential for word-of-mouth recommendations.

Promotion And Advertising

A successful turtle farming enterprise requires effective advertising and promotion. Consider the following strategies:

1. Social media channels may be used to display your tortoises. Engaging material, such as films and attractive photographs, may amass a sizable following.

2. Google Ads: Spend money on Google Ads to reach a larger online audience. Use tortoise-related terms, such as species and care.

3. Attend reptile expos and local pet fairs in your area. These events allow you to show off your tortoises to a specific audience.

4. Collaborations: For cross-promotion, team up with influencers or reptile specialists. Their support might increase your reputation and reach.

5. Build an email list of potential customers and send out newsletters with information on available tortoises, care recommendations, and conservation initiatives.

6. Conservation Initiatives: In your advertising, emphasize your dedication to conservation

activities. Buyers who wish to support ethical farming techniques may find this appealing.

To summarize, good tortoise sales require excellent marketing, advertising, and promotion. Building a good web presence, adhering to rules, and participating with the reptile community are all important aspects to ensuring the success of your tortoise farming business.

31

CHAPTER FOUR

Tortoise Farming Challenges And Problem Solving

Tortoise farming, while lucrative, is not without its difficulties. One of the most difficult challenges is to create an appropriate habitat that matches their native surroundings.

Maintaining the proper temperature, humidity, and shelter may be difficult, especially in climate-sensitive areas.

Tortoises are also prone to a variety of health problems, such as respiratory infections and shell rot. It is critical for their health that these issues are identified and treated as soon as possible.

Feeding can be difficult as different tortoise species have varied food requirements. Obesity and nutritional deficits can result from overfeeding or supplying the incorrect meals. To address these concerns, it is essential to conduct a study on the unique needs of the tortoise species you are raising and speak with qualified breeders or vets.

Problem-Solving Techniques

Effective problem-solving skills are vital in turtle farming when problems emerge. Regular health inspections and quarantine measures for new immigrants can aid in disease prevention. Maintaining a clean environment and giving enough nourishment helps lower the risk of common health issues. In times of disease, visit a reptile veterinarian who is familiar with tortoises.

To provide a stable environment for habitat-related difficulties, invest in high-quality enclosures and heating/cooling systems. Consultation with expert tortoise farmers or herpetologists might give useful ideas for dealing with certain challenges.

Managing Predators

Predators are a serious hazard to tortoises, especially hatchlings and youngsters. Birds, rats, and household pets are common predators. Use lockable containers with covers or wire mesh to restrict access to your tortoises. To dissuade prospective intruders, consider night-time security features such as motion-activated lights or alarms.

Natural landscaping items such as bushes and boulders can provide hiding locations and safety while rearing tortoises outside. Inspect enclosures regularly for evidence of predator activity and fix any breaches as soon as possible.

To preserve the well-being of these magnificent reptiles, successful tortoise farming necessitates a proactive approach to tackling issues, dedication to study, and the application of protective measures.

Future Tortoise Farming Trends

As we move forward, we should expect considerable changes in tortoise farming. One significant development is the incorporation of technology for precision farming. Advanced monitoring technologies, like as sensors and

Internet of Things devices, will be critical in assuring ideal circumstances for tortoise health and growth. Farmers will be able to remotely monitor aspects such as temperature, humidity, and feeding habits, providing a regulated and optimum environment.

Furthermore, conservation efforts will most certainly become more interwoven with tortoise farming. As many tortoise species risk extinction, farms may serve as conservation centers, raising and releasing endangered tortoises back into the wild. This dual-purpose strategy is in line with the rising worldwide understanding of the need for biodiversity preservation.

As a result of ethical concerns and a greater emphasis on animal welfare, the future of tortoise farming may turn toward more humane and natural breeding procedures. This might include developing conditions that are more similar to their native habitats, fostering better growth and behavior.

Finally, with a growing awareness of environmentally beneficial activities, tortoise farming may adopt eco-friendly projects. This might include using renewable energy, decreasing waste, and using sustainable farming practices to maintain the long-term viability and environmental responsibility of turtle farming.

Tortoise Farming Has Come To An End.

To summarize, turtle farming is a complicated and contentious activity that presents significant ethical and environmental concerns. While it may give economic benefits in some areas, it frequently includes unsustainable harvesting of natural populations, jeopardizing the extinction of several tortoise species.

The ethical consequences of keeping these slow-moving species in captivity cannot be overstated, since it frequently results in stress, health problems, and decreased well-being.

Conservation activities, like as breeding programs and habitat conservation, are critical for preserving tortoise populations and natural ecosystems.

Over exploitative agricultural methods, sustainable practices that value the well-being of these creatures and their environments should be supported. Finally, the survival of tortoise species is dependent on a balanced strategy that recognizes their inherent worth as well as their ecological importance.

To summarize, tortoise farming must be severely assessed, and efforts should be focused on the protection and preservation of these amazing species, both for their own sake and for the benefit of our planet's health.

www.ingramcontent.com/pod-product-compliance
Lightning Source LLC
Chambersburg PA
CBHW060012300526
45794CB00003B/1177